GW00361941

MINERAL WEALTH OF

SAUDI ARABIA

MINERAL WEALTH OF
SAUDI ARABIA

PHOTOS BY OCTAVE FARRA

TEXT BY CHRISTOPHER SPENCER

IMMEL
Publishing

**MINERAL WEALTH OF
SAUDI ARABIA**

IMMEL Publishing
Ely House, 37 Dover Street
London W1X 3RB

First published 1986.
© 1986 IMMEL Publishing

Phototypeset in Perpetua by
The Typesetting Company, London, England.
Printed and bound in Japan by
Dai Nippon Printing Company,
Tokyo.
Design and artwork by
Patrick O'Callaghan Graphic Design.

ISBN 0 907151 13 2

Page 2
Black sphalerite partially masked by
chalcopyrite in fine volcanic ash from
Khnaiguiyah.

CONTENTS

INTRODUCTION

Pyramid crystals of quartz from
Jabal Miskah

Saudi Arabia is a country known worldwide because of its oil riches, but how many people know that Saudi Arabia also contains the fabled Gold Mines of King Solomon (now known as Mahd adh Dhahab or the Cradle of Gold)? Even less well known are its deposits of Metallic Ore Minerals such as for iron, zinc and copper, and of Non-metallic Ore Minerals such as for phosphate, clay and glass sand. There are no known deposits of gemstone, but many of the common Rock Forming Minerals that are normally formless to the naked eye can be found as discreet and beautiful crystals in veins. It is a question of keeping one's eyes open and possibly learning the rudiments of geology.

Free gold in quartz from Shakhtaliyah

Pink granite from Al Jumum

Free gold in quartz from Shakhtaliyah

Petrified wood from near Riyadh

This book is a photographic record of some of the more interesting of the minerals, rocks and fossils that the authors have come across during their travels around Saudi Arabia. It is not intended to be a scientific work, although a thread of geological logic has been woven into the presentation, nor to be a comprehensive record, but only a collection of what any desert traveller may find in his wanderings in this vast and breathtaking land.

A SIMPLIFIED GEOLOGY OF SAUDI ARABIA

Photo by Christopher Spencer

Wadi in 'hard rock' landscape near Madinah. Note the flat surface of a young basalt flow in the hinterland

Travelling by road between Jeddah and Riyadh gives an excellent introduction to the varied geology of Saudi Arabia. One passes from the flat coastal plain abutting the Red Sea onto the 'basement' or the 'shield' of Western Arabia which is at its most majestic in the escarpment before Ta'if. This 'hard rock' domain is made up of usually dark-coloured, crystalline igneous and metamorphic rocks – the former originating directly from molten magma deep within the earth's crust and the latter being rocks that have been changed chemically/physically through enormous heat/pressure. From Ta'if to Ad Dawadimi the 'shield' is almost flat giving an impression of age. Then east of Ad Dawadimi, the landscape changes with much more mellow and golden colours. This is the 'soft rock' domain of the light-coloured, relatively soft, commonly friable, commonly bedded, sedimentary 'cover rocks' that originate either from the disintegration and redeposition of already existing rocks or from precipitation of carbonates and accumulations of animal shells. These are the rocks that make up Eastern and much of Northern Saudi Arabia, and they contain a record of the evolution of animal and plant life in the area over the last 500 million years.

Vast tracts of both the 'hard rock' and 'soft rock' domains are covered by much younger rocks. Ancient lava fields or 'harrats' that were spewed out between 20 million and 800 years ago form a north-south alignment in the west of the Kingdom. A trip from Jeddah to the city of Madinah by the splendid new Makkah-Madinah expressway passes through the hauntingly beautiful landscape of the Harrat Rahat which is dotted with conical volcanic mountains from whence gushed the hot lavas. The sedimentary 'soft rocks' are covered by extensive sand seas or 'nafuds', of which the Great Nafud in the north and the Rub' al Khali (Empty Quarter) in the southeast are the best known. These vast, almost impenetrable expanses of sand dunes are only a few thousand years old.

Thus Saudi Arabia can be divided into four main types of terrain – 'hard rock', 'soft rock', 'harrat' and 'nafud'.

A typical 'soft rock' landscape of limestone hills in Wadi Shadgam, near Hofuf

***Large crater filled with silts in the
basalt of the Harrat Rahat***

Photo by Christopher Spencer

An oasis in the dunes of the eastern
Rub' Al Khali
(Photo by Patrick Skipwith)

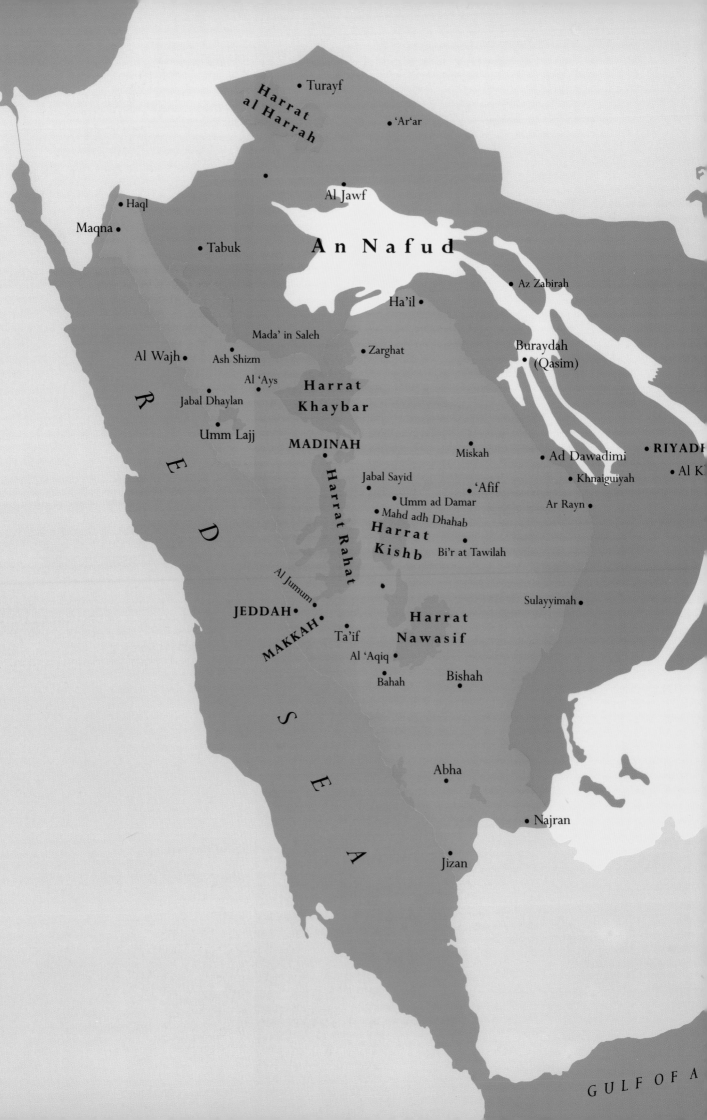

SIMPLIED GEOLOGIC MAP
OF SAUDI ARABIA

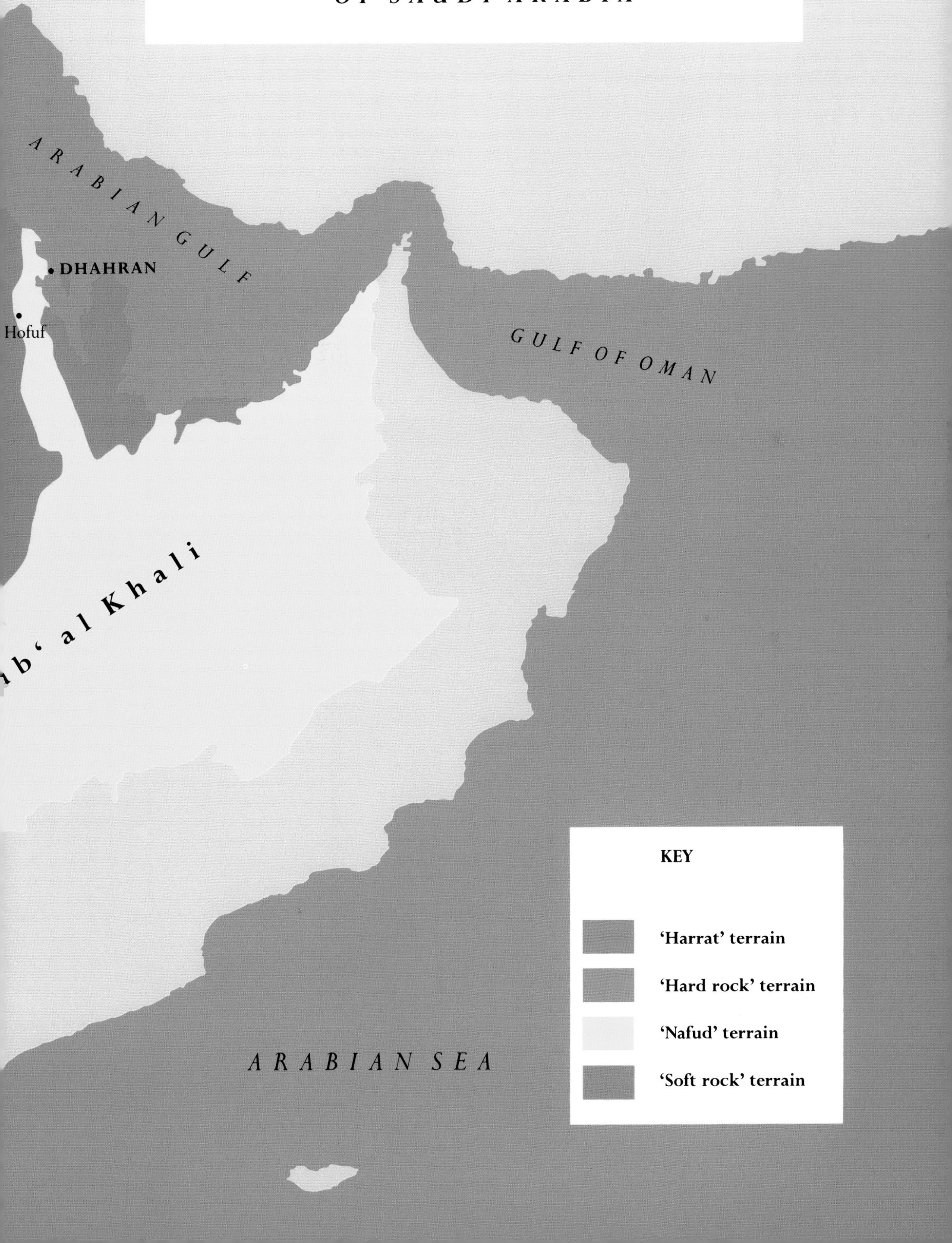

ARABIAN GULF

● **DHAHRAN**

● Hofuf

GULF OF OMAN

b' al Khali

ARABIAN SEA

KEY

'Harrat' terrain

'Hard rock' terrain

'Nafud' terrain

'Soft rock' terrain

'HARD ROCK' TERRAIN

The 'hard rock' terrain of Western Saudi Arabia is the dormant witness of some 700 million years during which there were at least three turbulent upheavals within the earth's crust, followed by long periods of relative quiet. From examining the rocks geologists can tell that 1200 million years ago, when the oldest rocks known in Saudi Arabia were formed, the planet Earth bore no trees and no soils; the only plants were lichens and moss-like algae, and the seas were inhabited by jellyfish and primitive soft-bodied creatures. The periods of turbulent upheaval saw a cycle that began with majestic volcanoes erupting stifling clouds of ash and cinder into the sea to form new, ephemeral masses of land that were often washed back into the ocean by wind, rain and waves to form thick deposits of sedimentary rock. Thundering earthquakes then gave rise to massive landslides both on the land and under water, and vast tracts of the earth's surface were hacked about, compressed, folded, eroded and buried. During this time, deep in the earth's crust and way below the belching volcanoes, extensive chambers of molten lava churned and heaved and rose into the crustal folds, there to be distilled and later cooled into coarse-grained granites and related gabbros. As remnants of the distillation process, hot fluids with concentrations of heavy metals (which originally made up only a very tiny part of the total volume of the molten lava) escaped to the surface through fractures and orifices, sometimes shedding their metal contents into the rocks through which they passed, sometimes cooling and solidifying with their metals caught up in the minerals that the geologist today seeks for a multitude of industrial and commercial uses.

These turbulent mountain-building phenomena accreted land to the huge southern continent of Gondwanaland from which the Arabian Peninsula was eventually, many eons later, to split away, and at the join of each accretion, mantle rocks from below the earth's crust would be squeezed to the surface. Eventually the turbulence of 700 million years ceased, the peaks and volcanoes were gradually eroded away to form a vast plain, and the next 480 million years were a time of quiescence during which the 'soft' sedimentary rocks were laid down over the now consolidated 'basement'. But recent earth movements associated with the opening of the Red Sea have lifted up this basement to a height of some 3000 meters and renewed erosion has cut out the present scarps, mountains and crags of the Asir that are so evocative of the turbulent age of volcanoes.

*The escarpment south of Bahah showing
the grandeur of the 'hard rock' terrain*

Crenulation folding in metamorphosed
20   *volcanic ash from Ash Shizm*

*Coarse-grained nepheline syenite
from Haql showing pink orthoclase,
black augite, and grey nepheline*

*Core from Khnaiguiyah showing
bedded volcanic ash that was laid
down under water*

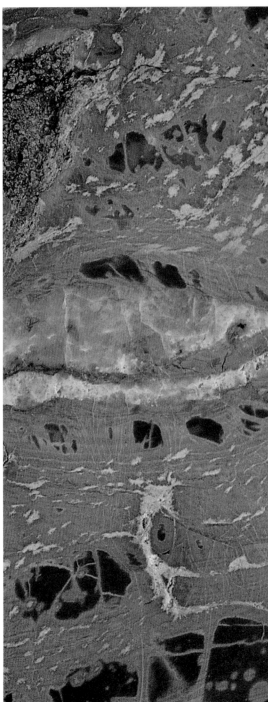

*Zinc ore in volcanic ash from a drill
hole at Khnaiguiyah. The zinc is in
the black sphalerite with pale
yellow pyritic stringers*

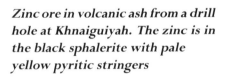

'Soft rock' terrain at Madain Saleh

'SOFT ROCK' TERRAIN

The 'soft' or sedimentary rocks of Eastern and Northern Saudi Arabia reflect a long period of geological calm. The vast plain of basement rocks was inundated and, for some 480 million years, formed a stable and shallow sea floor on which sand, shells and mud quietly accumulated. As the continent slowly subsided under the weight of the ever thickening pile of sands and muds, the earlier sediments became buried and compacted; the mud turned to shale, the sand to sandstone, and the shells to limestone.

Inselberg of sandstone at Madain Saleh

Then, about 20 million years ago, the area was gently lifted up and the solidified pile was exposed to be eroded into a landscape quite different from that of Western Arabia. Absent are the folds and the faults and the tortuous scars of mountain construction seen in the 'hard rock' domain. The resilient limestone and sandstone stand proud and untouched in massive, towering cliffs which, in some parts of the country, form an almost Manhattan-like skyline. The softer muds were blown away to leave long basins now filled with sand and sabkhah. In the Eastern Province the landscape is more subdued with only occasional 'inselbergs' (solitary towers of rock) on a flat, timeless plain.

Petrified wood reminiscent of delta swamps

Ripple marks preserved forever in sandstone near Ha'il recall the shallow seas of some 400 million years ago

Oil flare in the desert of the Eastern Province which hides huge reserves of oil

Although there were no hot concentrated fluids to deposit economic minerals in the 'soft' rocks, they are not without economic importance. On the contrary, it is these rocks that contain the oil wealth of Saudi Arabia; oil that originated from minute plankton-like organisms which abounded in the warm, languid, shallow seas. These shallow seas also, at some stage, saw the right conditions for the deposition of phosphate, whilst trees grew and soils formed on periodic delta-like structures to be preserved as lignite, bauxite, and fragments of petrified wood.

'HARRAT' TERRAIN

The Arabian Peninsula until some 20 million years ago was joined to the African continent and it is easily seen from the map that the opposing coastlines of the Red Sea are like two pieces of a jigsaw puzzle. The rift, which now forms the Red Sea, began to widen as Saudi Arabia pulled away from Africa and the first volcanoes for 500 million years began to erupt, for this was not a placid separation. Immeasurable volumes of lava, occasionally containing fragments of rock such as lherzolite and olivine cumulates (with their crystals of languid, limpid greens) from the very heart of the earth, were spewed out and flowed like thick treacle to form immense black basalt plateaux. The uninviting, almost lunar landscapes of the 'harrats' show vestiges of their more violent days with innumerable craters and cones now crumbling in the listless winds. But the upheaval is not over; the most recent eruption took place in historical times (6th century AH; 12th century CE) when the lava flow stopped just short of Madinah.

Ropy lava from a basalt flow near Madinah – a flow that stopped at gates of the city in historical times

A mantle inclusion of pyroxene from deep within the earth – lherzolite from Harrat al Harrah composed of black augite and reddish enstatite

Pale-green olivine cumulates from a mantle source caught up in a lump of basalt of the Harrat Rahat

29

Lava flowed from numerous volcanoes, filled valleys and covered lakes, to form extensive barren wastelands

'NAFUD' TERRAIN

A seemingly limitless sea of sand
in the eastern 'Al Khali

Photo by Patrick Skipwith

The great sand seas of the Rub' al Khali and the Great Nafud are extremely young in terms of the geological history of Saudi Arabia. Their curvilinear geometry, dramatized in the contrasting shadows of the scarlet setting desert sun, make these areas some of the most daunting and beautiful places on earth. That they are only recent is shown by an abundance of hand axes and other paraphernalia that can be found between the sand formations, indicating that man may well have been around before their formation.

Hand axes of rhyolite and andesite found between the dunes of the Great Nafud

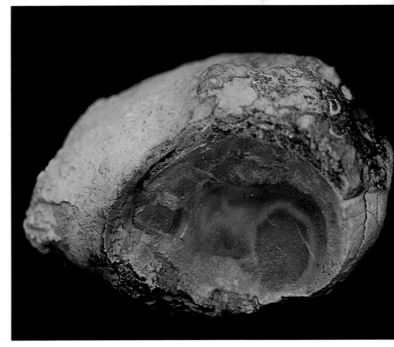

A prehistoric hammer head from near Al Jawf

ROCK FORMING MINERALS

QUARTZ

A cluster of well-formed quartz crystals from a vein in the 'hard rock' domain

Pegmatitic quartz from the 'hard rock' domain showing the typical pyramid crystal structure of this mineral

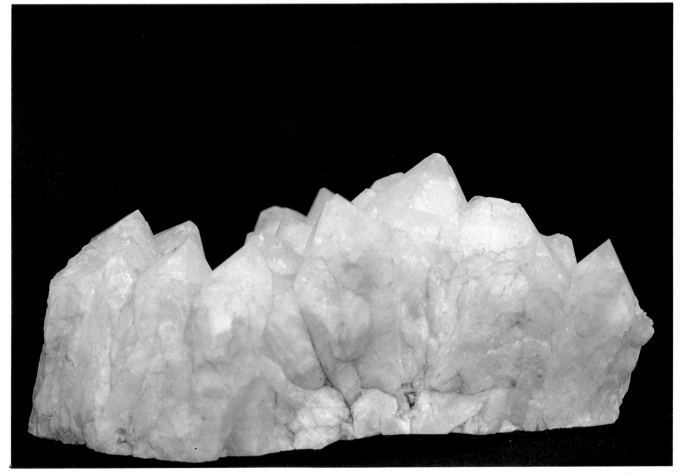

BIOTITE

A flake of black mica biotite such as
can be found in pegmatitic granites
in the east of the 'hard rock' domain

Although all minerals have specific crystal habits to which they must conform, to find well formed crystals is fairly rare. However, given the right circumstances, even the most humble of minerals such as QUARTZ and MICA can impart dazzling symmetries of crystalline form not generally associated with the formation of rocks. Also rare are the variant growths of a mineral; the most common of the minerals, FELDSPAR, is normally greyish-white to pinkish in colour, but can grow as the bright turquoise crystal AMAZONITE.

FELDSPAR

Well-formed orthoclase feldspar crystals surrounded by amorphous feldspar from a pegmatite at Jabal at Tuwalah, north of Madinah

AMAZONITE

A microcline feldspar from Awjah, near Makkah, with its character-istic brilliant turquoise colour

HORNBLENDE

The churning chambers of molten granite sometimes gave rise to curiosities such as the parallel crystals of HORNBLENDE from near Sawawin. Internal stresses within the earth's crust are thought to have caused crystals to grow in elongate aligned rods.

Rod-like hornblende crystals in a granite of pink feldspar and glassy quartz, from near Sawawin

GARNET

Cubes and rhomboid crystals of garnet encased in schist from near Ar Rayn

Certain minerals such as GARNET require high temperatures and pressures to form. These crystals seldom exceed a few millimeters, but occasionally they occur as small, rather resplendent crystals encased like some abandoned jewellery in the rock.

CALCITE

*Rhombs of vein calcite found
throughout the 'hard rock' domain*

Another mineral that may be found as collector's items in Western Saudi
Arabia is CALCITE, which is common in veins and fracture zones as well as
being the principal mineral of marble.

TOURMALINE

Typical three-sided prism of black tourmaline in quartz from Bi'r at Tawilah

TOURMALINE, a moderately rare mineral with a characteristic needle-like habit, is found with granites and may be associated with tin ores, topaz, wolframite and even kaolin. In some localities of the world it is cut as semi-precious stones for jewellery.

OLIVINE

*Green olivine crystals with black
pyroxene in a mantle inclusion
from the 'harrats'*

OLIVINE is an abundant mineral in certain types of dark, heavy rock such
as gabbro which form deep within the earth. Olivine is also found in the 'harrat'
terrain of Saudi Arabia and in particular in the volcanic cones where it
concentrates into nodules. These lie scattered on the surface awaiting a hammer
blow to reveal clusters of small, semi-transparent crystals, and some have been
known to produce small peridots, which is a semi-precious variety of olivine.

Peridot (semi-precious olivine) from Harrat Kishb

BERYL

BERYL is a rare mineral that contains the element berylium. The colour green is commonly associated with it, and rightly so, since in its most transparent but unfortunately rare forms it is better known as emerald. This specimen remains a curiosity, as no major deposits are known in Saudi Arabia, but in view of the vastness of the Kingdom one cannot rule out the existence of one or more worthwhile deposits.

Small blue-green beryl in amorphous pegmatitic quartz

MUSCOVITE

MUSCOVITE or 'white mica' is an exceptional mineral that crystallizes into 'books' or 'sheaves' of more than 30-40 cm across. Thin perfectly transparent sheets can be cleaved off with a sharp knife and used as a sort of glass. Because this transparent mineral does not melt even at very high temperatures, it was used as a porthole perspex in the furnaces of the steel, cement and refractory industries.

Laminated muscovite crystal with a pearly lustre from near Miskah

ZIRCON

ZIRCON, the principal mineral species of the element zirconium, is often associated with minute but measurable quantities of radioactive elements. From the ratios of radioactive elements weighed against the zirconium one can calculate the age of the rock since radioactive elements decay at fixed and known rates. In the barren, volcanic terrain of Western Arabia, much of the chronology of events has been based on such methods.

Brownish thorium-bearing zircon crystals from Ghurrayyah

SPINEL

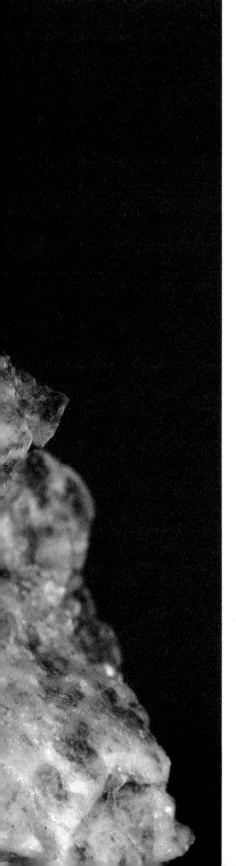

Black spinel from near Al 'Ays

Slices of deeper crust thrown up during the mountain building processes reveal, in the 'hard rock' terrain, rare species of minerals such as SPINEL. Transparent varieties have been taken for rubies in the past. However, spinels found hitherto in Saudi Arabia have proven mainly black, altered and of little to no gem value.

DESERT ROSE

PRECEDING PAGES
Gypsum crystals growing around
sand grains near Hofuf to form a
typical desert rose

THE DESERT ROSE or sand rose, exquisite, delicate, ephemeral, is indeed a rose in every sense of the term; it is one of nature's quirks and is quite unique to the desert. Its basic elements are crystals of gypsum that grow in the flat brine-rich sabkhahs so common in the Eastern Province. The gypsum crystals grow around and entrap windblown grains of sand at times when the water table is high. When the water table falls they become exposed due to wind deflation. Thus blooms the desert.

GEODES

Unopened cauliflower geode

Externally the GEODE comes in a variety of shapes and textures, but it is the interior that makes or breaks them for here in the void inward-growing crystals form a delicate array of spikes and facets. In Saudi Arabia the crystal linings are of colourless quartz, but in other areas of the world the crystals may be tinted purple, yellow, red or blue by minute quantities of metallic oxides.

GEODES

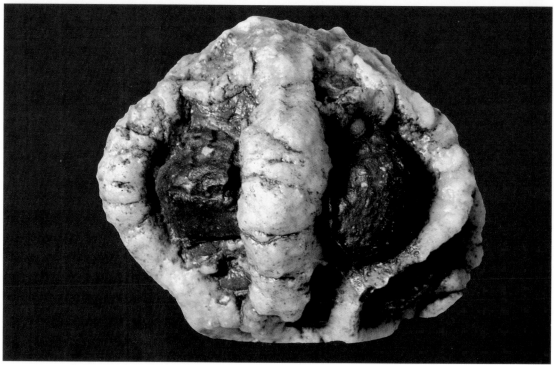

Unopened ribbed geode

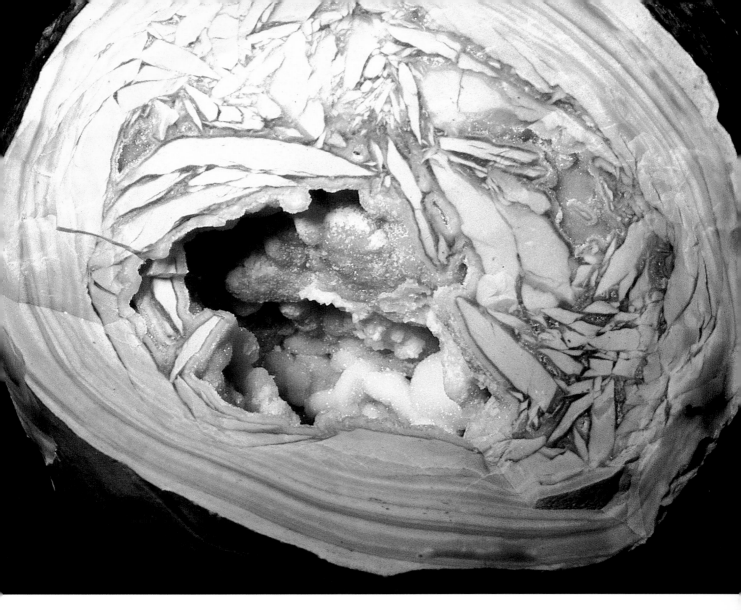

Smooth geode opened to show a microcrystalline quartz lining

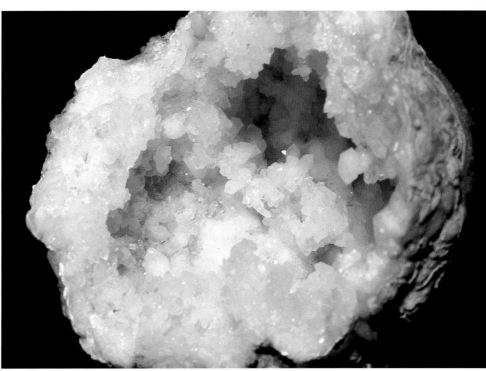

Cauliflower geode opened to show a lining of inward growing quartz crystals

The samples illustrated here come from near the Jordanian border in a bed comprised almost exclusively of geodes in Wadi al Ghinah, some 70 km southeast of Tabarjal; other locations exist north of Al Jawf and around 'Ar'ar.

METALLIC ORE MINERALS

Pure metals, excepting gold and platinum, seldom exist in nature; most metals occur in their more chemically stable oxides, sulphates, carbonates or chlorides and commonly in combination with other metals. To leave an iron bar out to rust in the rain for a few days is one way of seeing nature reverting the iron back to an ore. The ore minerals generally have a metallic glint to them, occurring in an almost endless variety of colours, many outstandingly attractive. Much of the 'hard rock' terrain of Western Saudi Arabia bears witness to man's endless search for metals; quartz veins that have been hewn out for their gold or copper; grinding stones in which vein material was crushed; smelting furnaces and piles of black slag often still tinted with the green of copper oxide. These activities took place during the Abbasid Khalifate in the 6th century AH (12th century CE), and the locations are still being investigated today to see what the ancients left behind.

PAGES 58/59
An aerial view of the Umm ad Damar ancient mine site, showing some of the 100,000 tonnes of slag heaps.

57

GRINDING STONES

Gold was freed from the quartz by grinding the ore in stones such as these

SLAG

Copper was smelted from the ore in primitive kilns, leaving a lava-like slag behind

SPHALERITE

ZINC is quite common in Saudi Arabia but is normally found only in modest quantities. The main zinc ore is SPHALERITE, closely associated with abundant pyrite and encased in ancient volcanic rocks. One very promising and newly discovered deposit is at Khnaiguiyah on the eastern side of the 'hard-rock' terrain.

Black sphalerite partially masked by chalcopyrite in fine volcanic ash from Khnaiguiyah

Black sphalerite associated with golden chalcopyrite in a quartz-rich volcanic rock from Khnaiguiyah

HEMIMORPHITE

Another variety of zinc occurs when it is oxidized in association with plenty of silica and water. Then the mineral species of HEMIMORPHITE is formed, which is used as the basic ingredient in calamine.

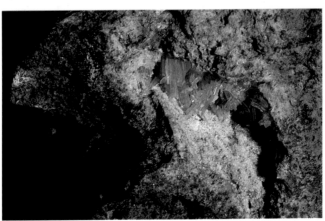

Transparent needles of hemimorphite in limestone from Jabal Dhaylan

AZURITE

Massive azurite associated with barites from Jabal Dhaylan

CHALCANTHITE

Blue crystals of chalcanthite from a weathered gossan zone at Jabal Sayid

A variety of COPPER ORES exist in nature with an almost awesome array of names: covellite, chalcocite, enargite, bornite, CHALCOPYRITE, cuprite, atacamite, MALACHITE, AZURITE, CHALCANTHITE, CHRYSOCOLLA, of which chalcopyrite is the most common ore in Saudi Arabia.

MALACHITE

Green malachite and blue azurite from Jabal Dhaylan

CHRYSOCOLLA

Deep blue, botryoidal chrysocolla from a weathered zone at Jabal Sayid

CHALCOPYRITE

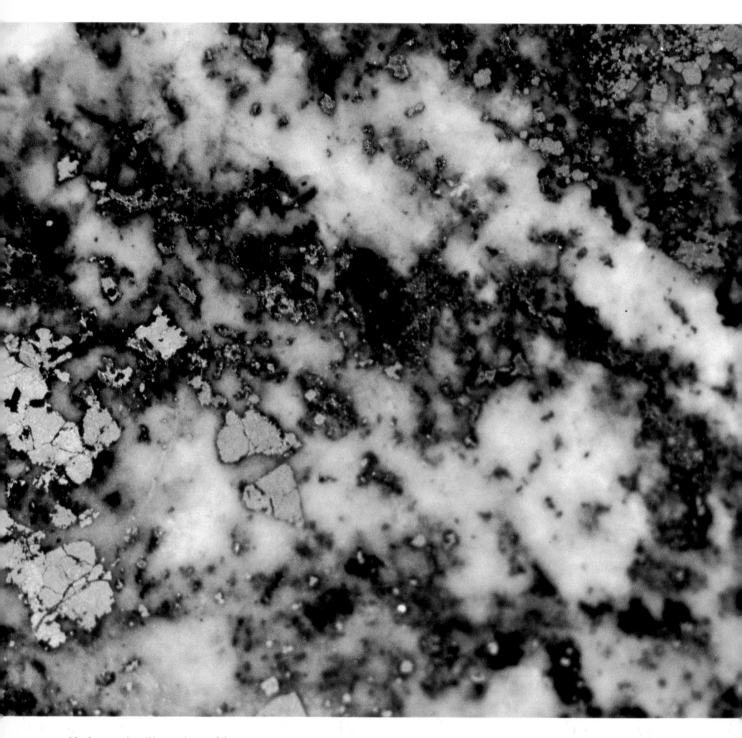

***Chalcopyrite disseminated in
quartz – a typical copper ore***

The main copper deposits of Saudi Arabia are at Jabal Sayid, 130 km south
of Al Madinah, and Al Masane, 35 km north of Najran. Both deposits are
pyritiferous, are associated with ancient volcanic rock, and are currently being
prepared for exploitation.

WOLFRAMITE

Black needle-like crystals of wolframite in quartz from Baid al Jimalah

Several small deposits of WOLFRAMITE (the main tungsten mineral) and SCHEELITE (which has the remarkable property of fluorescence in ultraviolet light) have been discovered in Saudi Arabia in recent years, with two important localities being at Baid al Jimalah and Bi'r at Tawilah in the central 'hard rock' area.

TUNGSTEN is very much a metal of the 20th century. It melts at 3410°C, is extremely hard as a pure metal and, when mixed with other base metals to form hard alloys, it passes on these qualities. Tungsten minerals such as WOLFRAMITE are normally associated with granites and often occur in quartz veins.

Massive wolframite in quartz from Bi'r at Tawilah

SCHEELITE

Scheelite in quartz from Bi'r at Tawilah

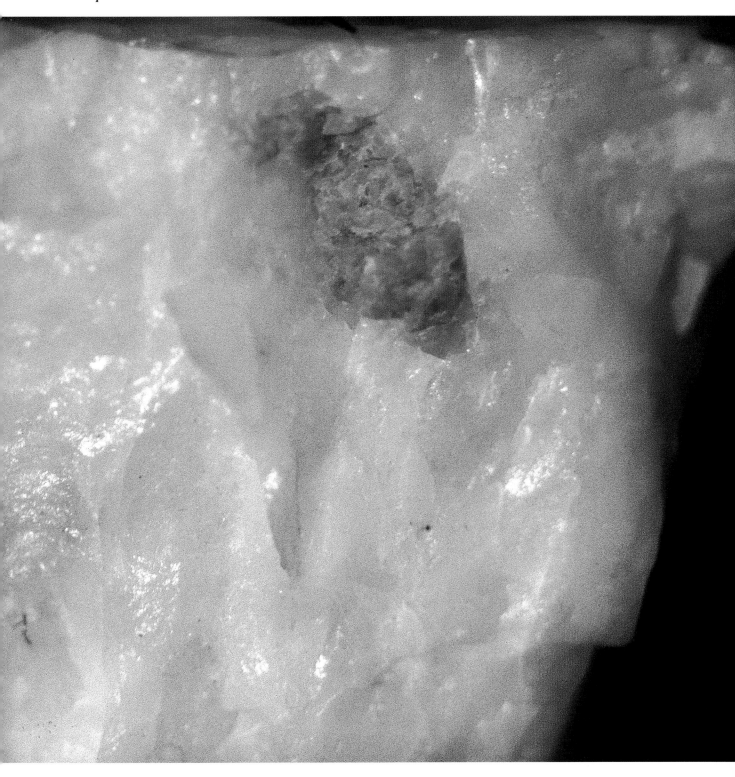

MOLYBDENITE

MOLYBDENITE, the sulphide of molybdenum, is another mineral associated with granites. It has proved rare in Saudi Arabia, where its main occurrence is with the wolframite at Baid al Jimalah. Molybdenum melts at 2500°C and is used as an alloy in special steels.

*Rosette of molybdenite in quartz
from Baid al Jimalah*

GALENA

Galena crystals from Jabal Dhaylan

The metal LEAD most commonly occurs as the sulphide GALENA and is found in association with silver and zinc. Galena has a metallic lustre, which must have made it attractive to primitive man, and is very soft and malleable which is probably why it was one of the first metals to be smelted. Although galena is moderately common in Saudi Arabia, no major economic deposits have yet been discovered.

GOLD

*Free gold cementing crystals of
quartz in a vein from Shakhtaliyah*

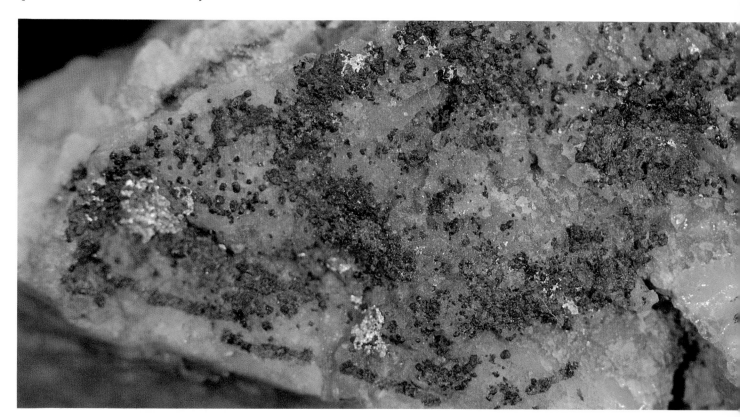

GOLD has always been one of man's most sought after treasures. It commonly occurs in its natural state, when it is very easy to extract from the encasing rock. Its natural softness as a metal and its low melting point allowed it to be fashioned into jewellery and ornaments with a minimum of tools. In Saudi Arabia over 400 ancient gold workings have been recorded, but they are mainly very small and no longer economical.

GOLD

The almost mythical King Solomon's gold mine at Mahd adh Dhahab was last worked in 1954, and it is possible that the mine will soon be opened up yet again. The gold occurs in association with the ubiquitous pyrite (or fool's gold) and is seldom in its free state; it is in a weak chemical liaison with silver in what are known as gold tellurides and requires a process using cyanides to extract it.

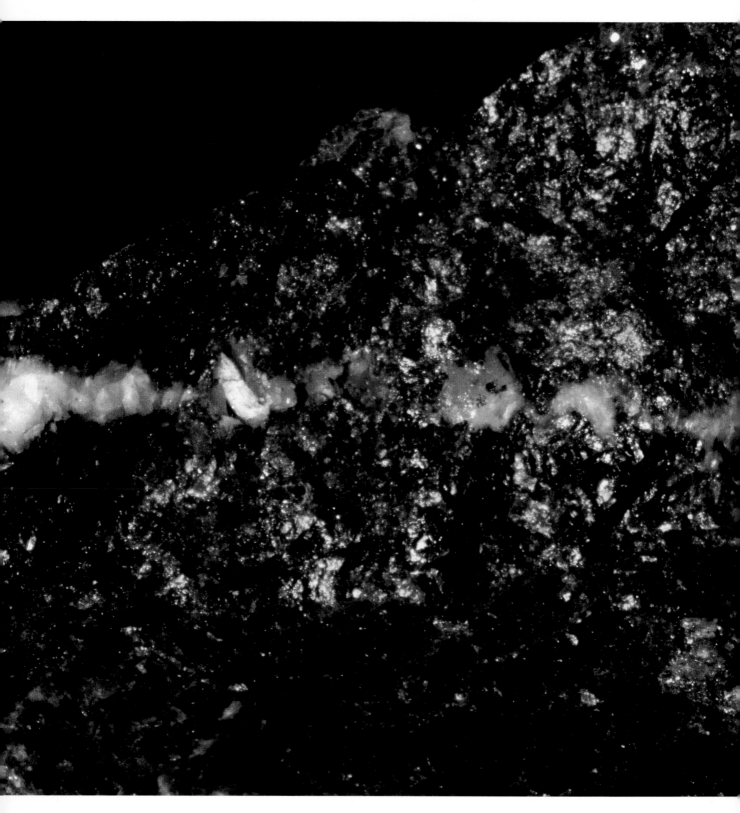

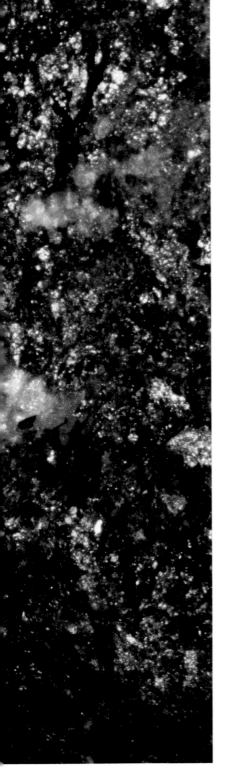

Gold ore from Mahd adh Dhahab is finely disseminated and cannot be seen with the naked eye. Minerals seen here are copper-coloured chalcopyrite and golden-coloured pyrite in quartz and rhyolite

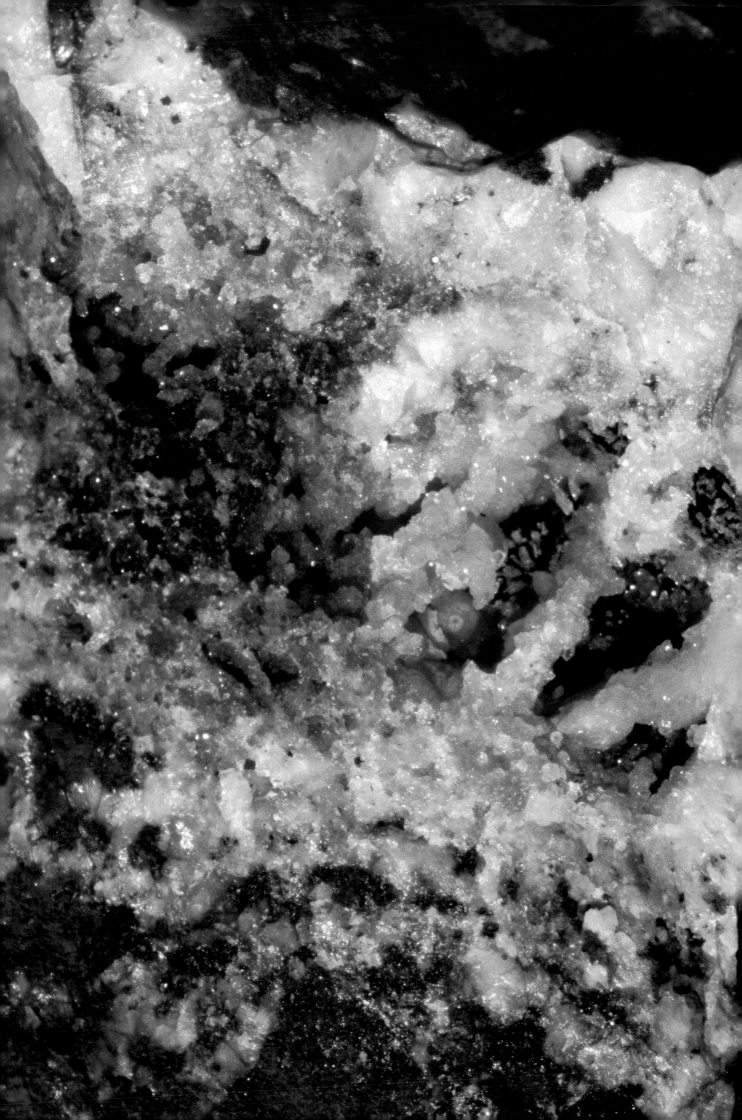

JASPILLITE

Above and Left:
Bands of iron-bearing jaspillite at
Wadi Sawawin

IRON, the most common of all the metals, occurs in a variety of ores. The major iron deposit in Saudi Arabia is at Sawawin, to the west of Tabuk. Here the iron-bearing rock is JASPILLITE, an extremely hard reddish mixture of silica and iron, which has a sedimentary origin associated with volcanic activity.

MAGNETITE

Another variety of iron ore is the iron oxide MAGNETITE of which the largest deposit is at Jabal Idsas in the east of the 'hard-rock' basement domain.

Massive magnetite from Jabal Idsas

PYRITE

The most common occurrence of iron is the sulphide PYRITE and its magnetic variety PYRRHOTITE. The largest pyrite deposit is in Wadi Wassat, near Najran. However, pyrite is also a very useful pathfinder mineral to the more precious metals of gold, lead, zinc, silver and copper with which it is commonly associated. Pyrite is also known as 'fool's gold' because of its yellow lustre and, to the untrained eye, its resemblance to gold.

Pyrite cubes (some oxidized black) in talc from near Ta'if

Massive pyrite (fool's gold) in quartz and basalt from Al 'Amar

Massive pyrite from Wadi Wassat

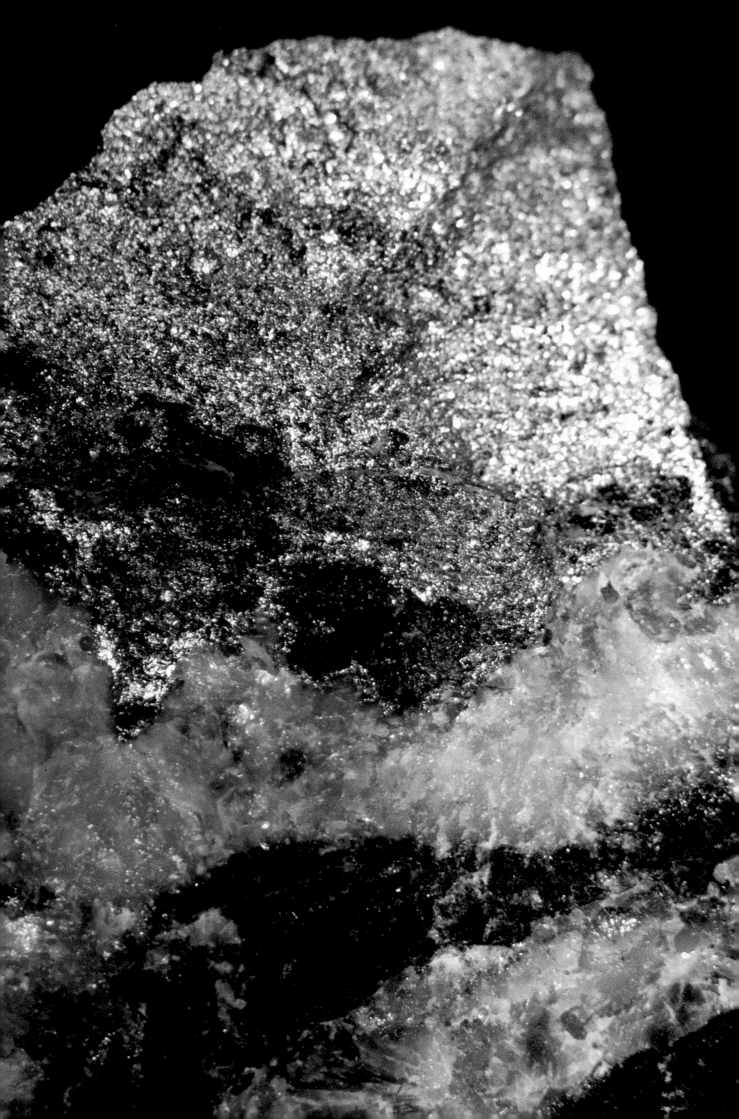

PYRRHOTITE

Pyrrhotite in volcanic rock

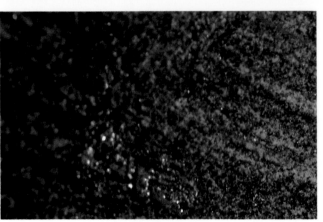

Oolitic iron ore from Wadi Fatimah   A deposit of iron in the form of HEMATITE is found near to Jeddah. This is the Wadi Fatimah oolitic iron ore deposit, whose origin is of quite a different nature – the iron precipitated in warm waters on small spheroids called ooliths or pisoliths.

PYROLUSITE

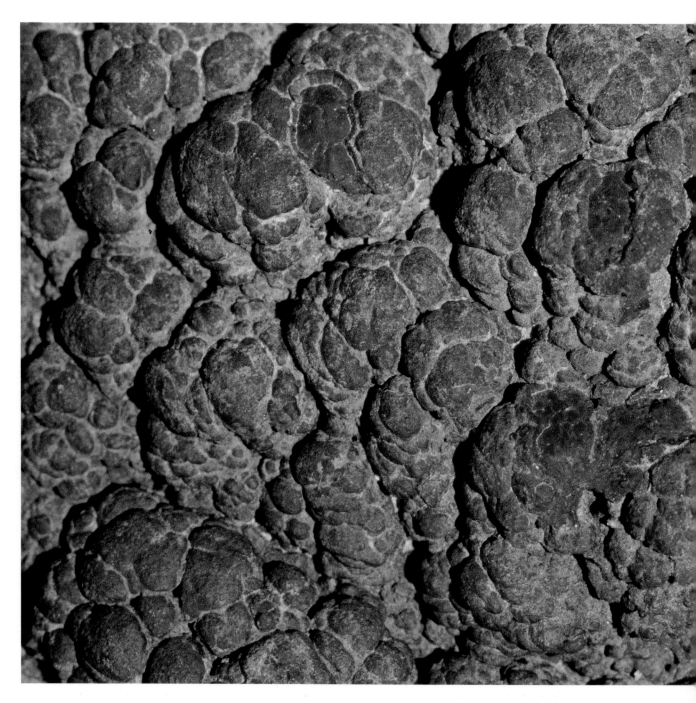

Reniform pyrolusite from Khnaiguiyah

MANGANESE is fairly ubiquitous in Saudi Arabia but is present only in very small quantities; thus its interest is not as an ore, but as fascinating minerals that occurt in a wide variety of structuring. Manganese muds are also forming today in the Atlantis II Deep, which lies 2200 m down at the bottom of the Red Sea and which one day could be mined. The main forms of manganese are as sulphides (albanite), hydroxides (PYROLUSITE) and manganite), carbonates (RHODOCHROSITE), and silicates (rhodonite).

RHODOCHROSITE

*Dendritic rhodochrosite in
limestone from Sulayyimah*

Massive chromite from near Al 'Ays

CHROMITE

The principal CHROME ORE is CHROMITE, which occurs in association with ophiolitic rocks that are sometimes thrown up from the mantle of the earth in the process of mountain building. Small chromite pods are associated with a slice of ophiolite near Al 'Ays, to the northwest of Madinah.

ILMENITE

ILMENITE, and associated rutile, are the most common sources of the element TITANIUM, a metal of quite outstanding qualities and much in demand in the aerospace industries.

*Massive ilmenite (black) with
disseminated pyrite (yellow) from
Jabal at Tuwalah*   89

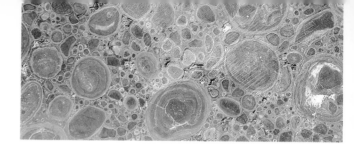

Pisolithic bauxite from Az Zabirah

BAUXITE

BAUXITE, the principal ore of ALUMINIUM (or aluminum), has only recently been discovered in Saudi Arabia. The deposit is at Az Zabirah a few hundred kilometers north of Qasim in the 'soft-rock' domain of eastern Arabia. Bauxite forms in areas of deep tropical weathering where heavy rains, luxuriant vegetation, and high temperatures cause the decay and weathering of aluminum-bearing rocks. This was clearly the case 150 million years ago with the Zabirah bauxite, which is associated with pisoliths (marble-sized concretions or spheroids).

NONMETALLIC ORE MINERALS

Gypsum wrapped around marls and shales near Maqna

The nonmetallic ores occur in both the 'hard' and 'soft' rock terrains and cover quite an extraordinary variety of rock types and deposit conditions. They range from crushed aggregate used in the construction industry, to diamond (not found in Saudi Arabia), a stone of pure carbon with a hardness unequalled in any other substance. Between lies a plethora of ores such as limestone for lime and cement, clay for bricks and ceramics, magnesite for furnace linings, quartz for glass, the silicon chip, and optical fibres, etc. Here we just have a fleeting glimpse of some of the more peculiar and beautiful ways these ores and substances can occur.

Red brick factory at Jeddah

GYPSUM

Transparent selenite from Dhahran

GYPSUM is a remarkable substance that is versatile and adaptable both in its uses and mineral habit, and Saudi Arabia is well endowed with deposits suitable for plaster production. The main centres all in the 'soft rock' domain, are at Maqna, Yanbu' and Jizan along the Red Sea coast, and at Buraydah, Al Kharj and Dhahran in the Eastern Province. The gypsum mainly occurs in its massive form, but SELENITE, a crystalline transparent variety, is also common in Saudi Arabia.

FLUORITE

FLUORITE is a very attractive mineral that is quite common in association with granites of the 'hard rock' terrain. It is much sought after by collectors because of its colour, which can vary from complete transparency to light blue or mauve, and because of its commonly well-formed cubic crystal habit. As an industrial mineral, fluorite is used in the steel industry as a flux and is a prime source for making the very powerful hydrofluoric acid that etches glass. In Saudi Arabia fluorite is found in one industrial-size deposit at Jabal Hadb Dahin near Mahd adh Dhahab, and as a mineral in many hundreds of small veins throughout the Kingdom's 'hard rock' terrain.

Cubic crystals of fluorite veining
amorphous fluorite from Jabal Hadb Dahin

Stalactitic baryte from Jabal Dhaylan

NON METALLIC ORE MINERALS

BARYTE

Crested baryte from Jabal Hamdhah near Maqna

BARYTES (or barite) occurs in veins and fractures, sometimes in association with fluorite, and in beds within limestone and gypsum. It is a very dense mineral and a lump in your hand along with a similar sized lump of quartz or calcite leaves no confusion as to which is the barytes. A beautiful but not a particularly common mode of the mineral is crested barytes. Barytes is used principally in drilling mud but is probably better known as 'barium meal' used in X-ray diagnostic techniques in hospitals. Only small deposits are known in Saudi Arabia, and these are found along the Red Sea coast.

ASBESTOS

ASBESTOS is a remarkable mineral that occurs as a matt of elongate flexible crystals, rather like coarse hair or matting, which can be spun into a cloth. Its extraordinary fire-resisting qualities are well known, but the fibres have proved dangerous to health for those that mine and work with the substance. In Saudi Arabia, asbestos in the form of CHRYSOTILE is best known in veins associated with ophiolitic rocks, such as to the west of Al 'Ays.

Cross-fibre chrysotile (asbestos) in serpentinite from near Al 'Ays

Long-fibre chrysotile (asbestos) from near Ha'il

MAGNESITE

The 'cauliflower' habit of MAGNESITE and its brilliant white colour makes this mineral stand out in a desert landscape. In Saudi Arabia it occurs in veins and fractures associated with ophiolite belts, and in two large deposits at Zarghat and Jabal ar Rokham where it has replaced ancient limestone. As magnesite has a very high melting point, its most important use is in refractory bricks and furnace linings.

Typical cauliflower habit of
magnesite from Zarghat

PHOSPHORITE

PHOSPHATE as a commercial substance is pretty much confined to the 'soft rock' terrain in the far north of Saudi Arabia. Here extensive deposits of PHOSPHORITE exist at Al Jalamid, Turayf and Tabarjal, under the endless plains which stretch north into Jordan and Iraq. The ore is characterized by small white 'pellets', rather like grains, amassed within generally grey or black limestone or chert; the material is seldom of aesthetic value. The phosphate derives from fish and other organisms concentrated in certain well-defined zones on shallow sea floors.

Phosphate rock from Tabarjal

KYANITE

Kyanite (blue) in quartz (transparent)
and feldspar (brown) from Jabal Kirsh

KYANITE is an exclusive 'hard rock' metamorphic mineral formed under 'high pressure' conditions; its presence indicates the very heart of mountain building processes. Once seen, kyanite is rarely mistaken for any other substance, due to its stunning light-blue colour, and in Saudi Arabia a whole mountain of the mineral has been discovered at Jabal Kirsh south of 'Afif. Uses of kyanite are for the most part refractory, in ceramics and other processes where very high temperatures are encountered.

GRAPHITE

Massive graphite from Wadi Bidah

GRAPHITE is found in moderately metamorphosed 'hard rock' terrain, but no commercial deposits have yet been discovered in Saudi Arabia. Like diamond, it is 100 percent carbon, but whereas diamond is the hardest of the minerals, graphite is one of the softest – it is used for the 'lead' of pencils and as a component in lubricants.

Massive halite, showing cubic symmetry, from the Jizan salt dome

SALT

Ancient SALT deposits entrapped as thick beds in the sediment show a phenomenon unique to evaporite deposits. This is the ability to flow rather like a highly viscous liquid and (in flowing) movement tends to be upward rather like a mushroom displacing surrounding strata and forming domes (or diapirs) such as that on which the town of Jizan is built. This process takes many thousands of years, but the halite crystals still retain their cubic symmetry.

Pure diatomite as is found in places within the Great Nafud

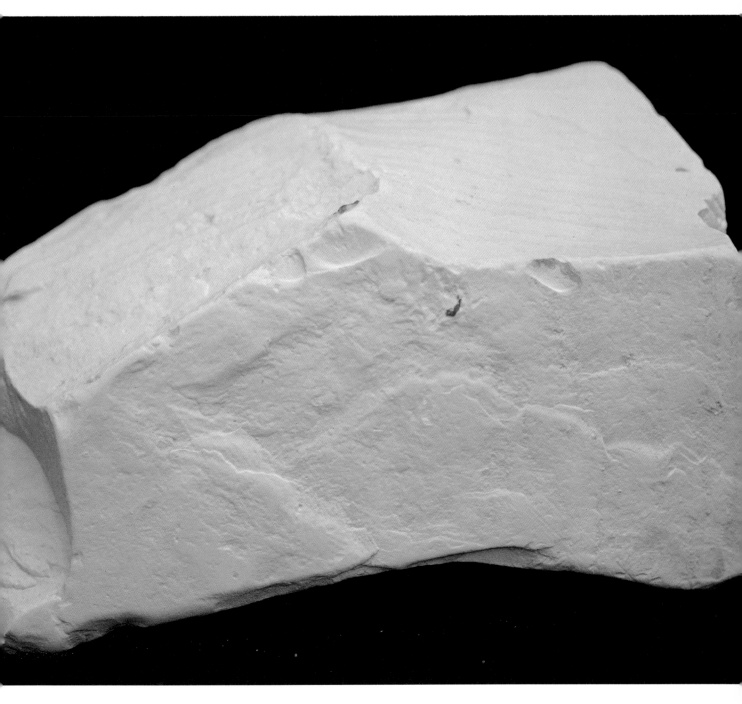

DIATOMITE

DIATOMITE is a rock composed of millions of minute skeletons of microscopic animals known as diatoms which inhabited still ponds and lakes. With its unique, highly porous structure, diatomite is used as a filter in various industries, as a filler or additive to paints, plasters, and as a mild abrasive and catalyst carrier. A deposit has been discovered in Saudi Arabia between the dunes in the Great Nafud northwest of Ha'il, and its commercial potential is still being assessed.

FOSSILS

112

The sedimentary rocks of eastern and northern Arabia were laid down during the period of the earth's history that saw animal and plant life develop and abound. The record of this life can be read from the fossils (the traces of animal or plant life preserved as skeletons, moulds, trails, etc.) found in the sediments that buried and preserved them. The examples presented here are amongst what we could call the 'easy to find' groups; all are fairly abundant at the localities where they are found. They are also presented in a vague chronological order of appearance in the fossil record.

PRECEDING PAGE
Crocodiles in a swamp
Illustration by Stephen Lee

Ammonites in water
Illustration by Stephen Lee

WORM TRAILS

On the muddy tranquil seabed, some 400 million years ago, unknown creatures possibly worms or crustacea, filtered the sediment for food particles. Their lifestyle was simple and monotonous but their TRAILS are encased forever.

*Worm? casts preserved in sandstone
near Tabuk*

GRAPTOLITES

GRAPTOLITES are amongst the most enigmatic of organisms found anywhere. They clearly floated around the oceans of 400 million years ago but what the soft parts looked like remains a mystery. They lived during a very specific period of time in geological history and, because they occur all over the world, it is possible to correlate rocks which contain them over enormous distances.

The 'tuning fork' graptolite in shale from near Ha'il

AMMONITES

Whilst dinosaurs roamed the earth some 150 million years ago, the tepid seas abounded with squid-like creatures in their coiled shells – the AMMONITES. This group of animals, of which the present day 'Nautilus' is the lone survivor, were plainly successful during a long spell of earth's history. Ammonites were very widespread; they can be found on every continent and scientists have concluded that, like graptolites millions of years before them, they lived at the sea surface, roaming the oceans over vast distances.

Ammonite in limestone from near Dhruma

Silicified clams from near Turayf

Ribbed oyster from near Turayf

OYSTERS AND CLAMS

OYSTERS AND CLAMS have assured their preservation in the fossil record because of their hard shells as opposed to soft-bodied creatures such as jellyfish which are extremely rarely preserved. They are abundant in the far north of Saudi Arabia in association with the phosphate-bearing rocks, where the ribbed oyster is found preserved directly in the chalky limestone and where the clams have been preserved through silicification.

117

NUMMULITES

NUMMULITES, which resemble coins in shape and size, are not well known and are confined to strata of a specific age. Banks of these creatures formed in reef-like structures such as are now exposed in the phosphate country between Tabarjal and Turayf. It is surmised that amoeba-like creatures lived within myriads of little chambers inside the shell.

Nummulites in limestone east of Tabarjal

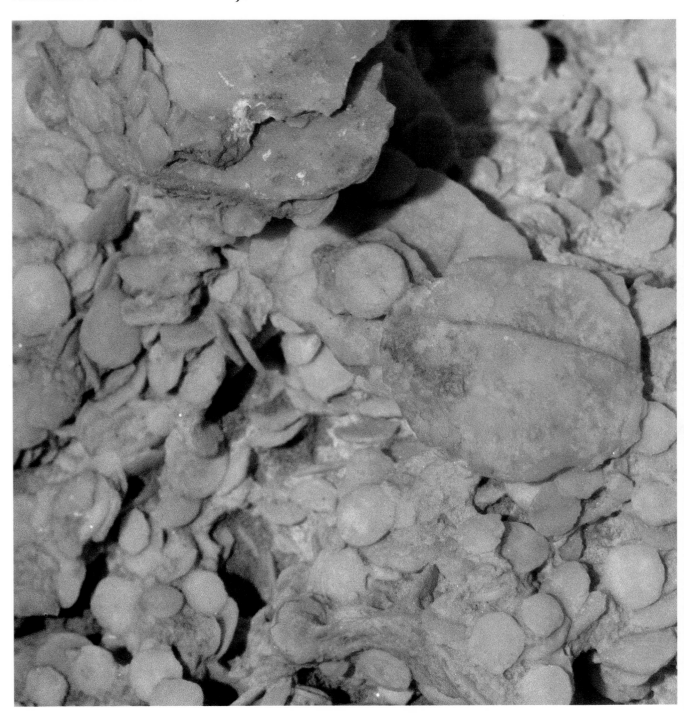

PETRIFIED FOSSIL WOOD

Substantial vegetal fragments have been preserved in some localities in the Kingdom, but best preserved are the tree-like plants and, generally speaking, the bigger the better; hence tree trunks are quite common and leaf imprints are quite rare. PETRIFIED WOOD in Saudi Arabia, dating back some 100 to 200 million years, is well known in the Qasim and Riyadh areas, with the petrification process (replacing original organic material by silica) often so gentle that the wood's complete structure is preserved. More fragile plant fragments can be found near Jeddah in the iron-oolite-bearing sediments of Wadi Fatimah.

Section of petrified tree trunk from Qasim showing growth rings

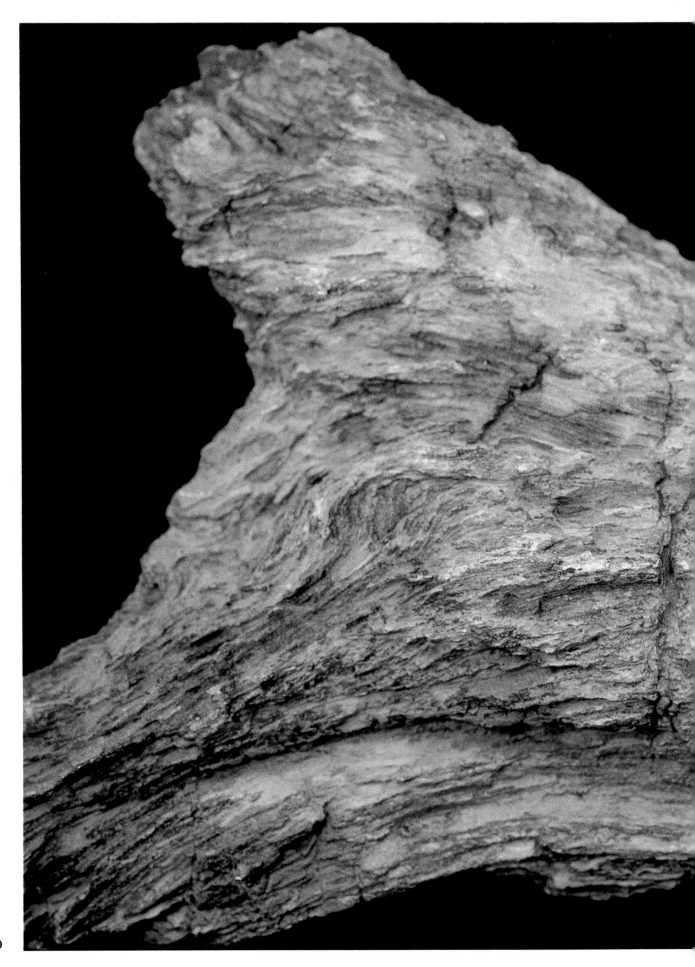

Petrified branch from east of Riyadh

Lower jawbone of a crocodile-like creature from east of Ta'if

CROCODILES

Thriving colonies of water living amphibians existed some 55 million years ago just 20 km east of Ta'if, in lakes and swamps now covered by the majestic black Harrat Hadan lava field. Vertebrae of some fairly substantial creatures closely related to the CROCODILES of today are to be found in the light-coloured beds of limestone and claystone beneath the basalt. Other vertebrate fauna (ancient mammals, lizards and fish) recorded from Saudi Arabia include sharks near Tabarjal, and elephants, hippopotami and rodents from Hofuf and Jizan.

Vertebra of a crocodile-like
creature from east of Ta'if

ACKNOWLEDGEMENTS

To be rambling in the wild in communion with Nature has always been the most relaxing part of my leisure hours. The panoramas in the mountains or the desert or along the sea shore can be breathtaking, but equally fascinating for me are the details that make up the environment – the birds, the plants and the rocks.

Many stones and rocks have shapes, colours and textures that make them works of art and, for many years, I collected them without ever trying to understand the mystery of their creation. The superficiality of my interest ended in Jeddah when a geologist friend told me how my stones were formed, what types of rock they were, and which ones were of mineralogical interest and why. He then explained to me the rudiments of geology and rock collecting, and introduced me to the geological hammer and the x5 magnifying lense, I was fascinated. On my next trip into the desert I began to see a new world imprisoned in the heart of the stone – it was no longer merely a stone with a beautiful shape, but a stone that was alive and with a story to tell of the world as it was tens or hundreds of millions of years ago.

I developed a particular enthusiasm for minerals when, with my family in the Taif mountains, we discovered a white stone that contained a perfectly formed mineral; a mineral that was yellow; golden yellow. I had sufficient knowledge by then to know that the stone was quartz and probably came from a vein and that veins were places where one could find gold. We all became very excited and spent the rest of the day turning over every white stone in the vicinity, many of which also contained the yellow mineral. Our imaginations started to run wild – had we discovered a gold deposit? would we now become wealthy? First, however, we would need some expert advice on how rich the vein was likely to be, and that evening I took our specimens to my geologist friend and described the environment in which we had found them. He listened carefully to my story, examined my samples and smiled. "Yes" he said "it is a gold", and my heart began to beat wildly "But", he continued "only Fool's gold and just about worthless". He explained that it was a mineral called pyrite that has been given the Fool's gold because for centuries it has deceived people like me; then and there I resolved to find out more about minerals.

Since that day, several years ago, I have travelled extensively in Saudi Arabia, taking every opportunity to add to my collection of local rock types, minerals and fossils. I also began photographing many of the specimens so as to be able to share my pleasure with friends, and it was my collection of photographs that gave me the idea for the present book Christopher Spencer ▷   **127** agreed to write a text, but the book would still not have been possible without help from many other geologist friends such as M. Kluyver, C. Thorber, H. VanDaalhof, M.M. Mawad, J.F. Labbé, R. Vazquez-Lopez, P. Skipwith, and R. Yared, who lent me further specimens for photography and gave me fruitful advice on the mineral wealth of Saudi Arabia. I should also like to make a special acknowledgement to my dearest wife, Leyla, who helped in the selection of photographs to be used, and to Dr. Skipwith who edited and helped format the book.

OCTAVE FARRA

France, November 1986